BOOK II
ALGEBRAIC SUDOKU

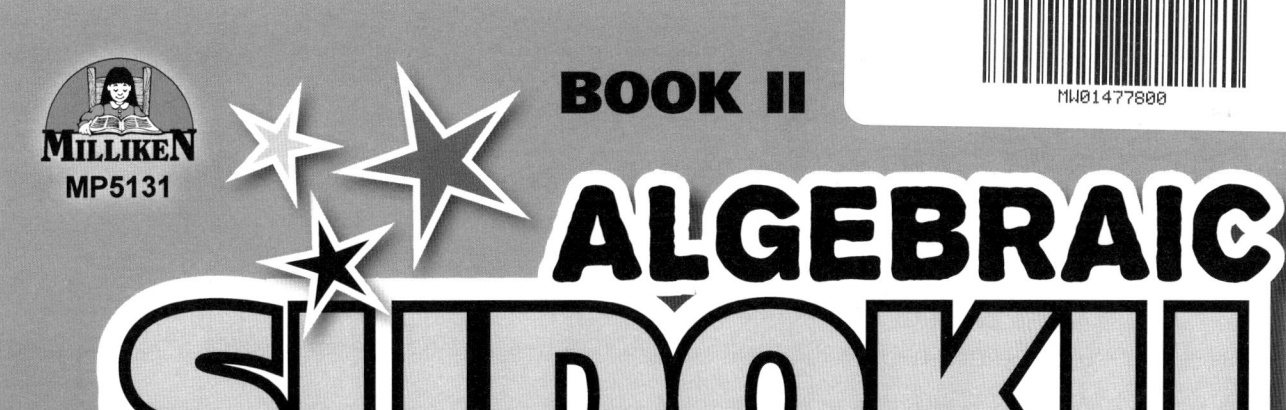

MP5131

A Fun Way to Develop, Enhance, and Review Students' Algebraic Skills

Author: Tony G. Williams, Ed.D.
Editor: Howard Brenner
Design and Layout: Kati Baker

Copyright: 2011 Lorenz Educational Press, a Lorenz company, and its licensors.
All rights reserved.

Permission to photocopy the student activities in this book is hereby granted to one teacher as part of the purchase price. This permission may only be used to provide copies for this teacher's specific classroom setting. This permission may not be transferred, sold, or given to any additional or subsequent user of this product. Thank you for respecting copyright laws.

Printed in the United States of America

ISBN 978-1-4291-2270-2

MILLIKEN
P.O. Box 802 • Dayton, OH 45401
www.LorenzEducationalPress.com

Dear Teacher or Parent,

Welcome to most exciting and beneficial algebraic learning resource on the market today. *Algebraic Sudoku* will teach your child/student algebra in a way never previously accomplished. In this book, the second part in a series of two, you will find 33 Sudoku puzzles, each of which aligns with the standard algebra curriculum. This book is divided in five units:

- Unit I: Systems of Equations
- Unit II: Polynomials
- Unit III: Factoring
- Unit IV: Radicals
- Unit V: Quadratic Equations

Each Sudoku puzzle is like a mini-lesson, with background, discussion, strategy, and demonstration for solving each problem. By solving the algebra problems, students are given enough data that will allow them to reason their way through the remaining cells of the Sudoku puzzle that follows. Each Sudoku puzzle includes a solution key. Every puzzle, along with the problems and mini-lessons associated with it, are presented on a ready-to-use, reproducible master that can be easily copied as transparencies for full-class instruction and discussion.

Algebraic Sudoku is an ideal supplement for students currently enrolled in algebra or students planning to take algebra in the near future. It is an excellent tool for kids and grown-ups to keep their algebraic skills sharp after having completed an algebra course. It is wonderful for home-schoolers as well as students in a traditional classroom. It makes an excellent companion during the summer break. Not only is *Algebraic Sudoku* fun and exciting, it challenges students' minds with exhilarating puzzles that develop logic, reasoning skills, concentration, and confidence.

I hope that you find the *Algebraic Sudoku* puzzles, and the mini-lessons and problems associated with each, the most fun and exciting way to achieve algebra mastery.

Sincerely,

Tony G. Williams, Ed.D.

Dedicated to my parents, Jean and Grady, who are retired math teachers, and to my wife Sharon and our triplets TLC (Tony, Leah, and Christina)

Also, to my brilliant Uncle Charles, and my siblings, Cheryl and Bruce, two committed math educators

TABLE OF CONTENTS

Introduction 4
How It All Works 5
Suggested Strategies 6
Pencil Marks.. 7

Unit I: Systems of Equations 10
Unit II: Polynomials 18
Unit III: Factoring 25
Unit IV: Radicals 33
Unit V: Quadratic Equations... 39

Answer Key 43

INTRODUCTION

The word *Sudoku* is Japanese—*su* means *number*, and *doku* refers to a single place on a puzzle board. Although its name is Japanese, the game origins are actually European and American. Sudoku puzzles come in a variety of sizes or grids (9x9, 6x6, 5x5, 4x4, etc.). The objective is to fill the grid with digits so that each column, row, and each of the nine 3×3 sub-grids that compose the puzzle contain all of the digits from 1 to 9. (We will discuss strategy later in this section.) Most of the Sudoku puzzles in the book are of the 6x6 variety, with a few 9x9 and 4x4 grids added to the mix.

Today, Sudoku is played all over the world. Many newspapers have added the game alongside the crossword puzzle and, as a result, have reported a jump in circulation. *USA Today's* most recent list of best-selling books tracked that seven of the top 100 were compilations of Sudoku puzzles. Also, millions of dollars are being made from Sudoku software, games, and online programs. But why? The most obvious reason is that Sudoku is fun; challenging, but fun! There is a sense of accomplishment one receives when successfully completing a puzzle. Sudoku tests one's ability to concentrate, reason, and think logically.

When you combine the excitement of Sudoku with the importance of algebra, you have a winning combination. Students who successfully complete an algebra course are twice as likely to graduate from college as students who lack such preparation (Adelman, 1999; Evan, Gray, and Olchefske, 2006). The majority of employees who earn more than $40,000 a year completed algebra in high school (Achieve, Inc., 2006). A national poll revealed that two-thirds of the students who completed algebra were well-prepared for demands of the workplace (Carnevale and Desrochers, 2003). And yet, there are increasing numbers of students who are not prepared for and fail to successfully complete algebra, as is evident by the vast and growing demand for remedial mathematics education courses among students in four-year colleges and community colleges across the nation. Data shows that 71% of America's degree-granting institutions offer an average of 2.5 remedial courses for skill-deficient students (Business Higher Education Forum, 2005). Overall, these deficiencies are further intensified by factors such as income and race. Research shows that most children from low-income backgrounds enter school with far less knowledge than peers from middle-income backgrounds, and the achievement gap in mathematical knowledge progressively widens throughout their PreK-12 years (National Mathematics Achievement Panel, 2008). However, these achievement gaps can be significantly reduced or even eliminated if low-income and minority students increase their success in high school mathematics and science courses (Evan, Gray, and Olchefske, 2006).

Algebra is essentially both the bridge and the gateway for many students. Algebra can lead students into an exciting world with opportunities awaiting them. We, as math educators, must be as innovative as possible in reaching our students, in enabling our students to reach their fullest potential. I hope this book will be a valuable resource as you strive for success in the classroom.

HOW IT ALL WORKS

This book is divided into three units, with each unit divided into various content areas. For each Sudoku puzzle, there is a specific content area that often includes a mini-lesson that describes the concept and provides an example(s).

The Basics
For each lesson, students must first complete the assigned algebra problems. They will then be asked to either place their answers directly into the Sudoku puzzle, or match their answers and place the corresponding numbers in the Sudoku puzzle. From their efforts, students will be given enough numbers to begin working on the Sudoku puzzle. It is then up to the students to solve the remaining cells of the puzzle.

A system has been devised to easily identify specific cells of a puzzle, using alphabets horizontally across the top and numbers vertically along the left side. Here are some examples:

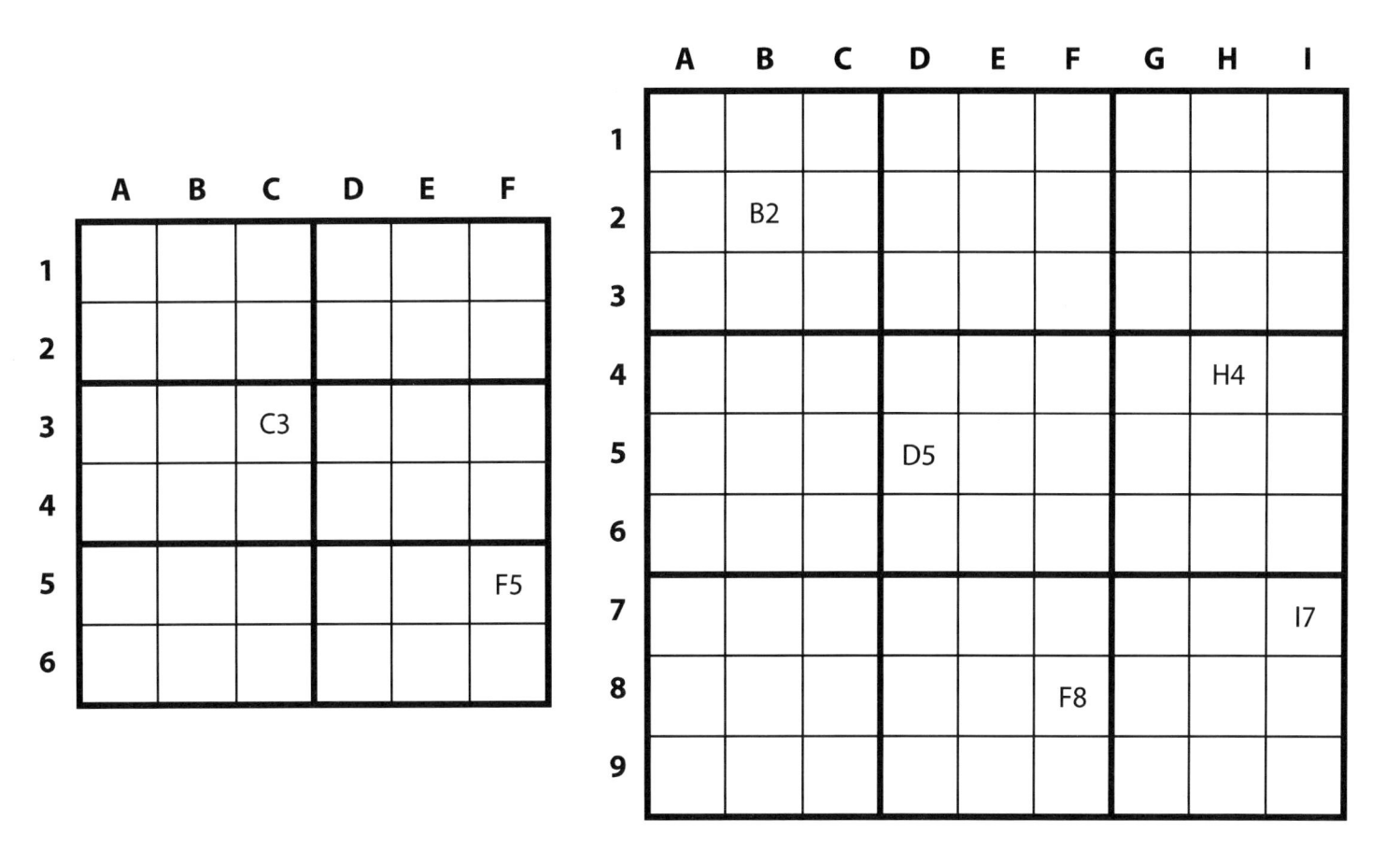

Solutions are provided for each Sudoku puzzle. Whenever possible, classroom discussion should be held to address any remaining questions and to provide further clarification.

Negative Numbers
It is important to note that the cells in these Sudoku puzzles may have negative numbers. Answers with incorrect signs are considered wrong. However, for the purpose of completing the entire grid, a negative number represents a positive number in the counting sequence. For example, a "-2" represents "2". In other words, it is okay to have a sequence such as "-1, 2, -3, 4 -5, 6, -7, -8, 9", provided that these numbers are the correct solutions to the problems presented in the lesson.

5

SUGGESTED STRATEGIES

Solving Sudoku puzzles should rely more on logic and reasoning, rather than guessing, arithmetic, or any mechanical system. For the novice, it will take some practice and specific strategies; however, the process can be mastered rather easily over a short period of time.

Sudoku Basics
The rules of Sudoku dictate that every cell in the grid is filled with a number. For a 4x4, you would use the numbers 1 to 4; for a 6x6 grid, you would use the numbers 1 to 6; for a 9x9 grid, you would use the numbers 1 to 9; and so on. The restriction is that you can only use each number once in each row, each column, and in each of the box-shaped sub-grids (indicated by bold lines). Also, for the purpose of *Algebraic Sudoku*, a negative number represents a positive number in the counting sequence.

Here are some typical examples of some completed Sudoku puzzles.

	A	B	C	D
1	-1	2	4	3
2	3	-4	-2	1
3	2	1	-3	-4
4	4	3	1	2

	A	B	C	D	E	F
1	4	3	-6	2	-5	1
2	1	-5	2	4	6	-3
3	6	-2	1	5	3	-4
4	5	4	-3	6	1	2
5	-3	-6	4	-1	2	5
6	2	1	5	3	4	-6

	A	B	C	D	E	F	G	H	I
1	1	4	-9	3	-5	2	6	7	8
2	-2	-6	3	1	-7	-8	9	-5	-4
3	-7	8	-5	6	4	9	-3	2	1
4	3	2	-1	9	-8	4	5	6	7
5	6	-7	8	5	2	3	-1	4	-9
6	-5	9	-4	7	-6	1	8	-3	2
7	8	-5	6	-2	9	-7	4	1	-3
8	4	-1	2	8	-3	6	-7	-9	5
9	-9	3	7	4	1	5	2	8	-6

PENCIL MARKS

One of the best strategies to develop students' logic and reasoning skills in solving Sudoku puzzles is by using pencil marks. (Very few players/students can complete Sudoku puzzles by only writing in the final numbers.)

The pencil mark method is an excellent way to get started. Students can learn various shortcuts, patterns, and even their own methods as they advance in their analysis. Let's do an example of the pencil mark method using a 6x6 puzzle. Initially, you will be given the numbers of a few cells to get you started:

	A	B	C	D	E	F
1						
2		5	2			3
3	6				3	
4		4				2
5	3			1	2	
6						

Step 1: First, write in all the possible numbers for the vacant cells. When looking for possible numbers for a vacant cell, you must first look at the given numbers in the same line as that cell both horizontally and vertically, as well as the given numbers in the same sub-grid. For example, for cell A1, vertically we have a 6 and a 3, horizontally there are no numbers in the row, and in its sub-grid we have a 5 and a 2. Therefore, the possible numbers for cell A1 are 1 and 4.

	A	B	C	D	E	F
1	1 4					
2		5	2			3
3	6				3	
4		4				2
5	3			1	2	
6						

7

Initially, it is recommended that you do this for every vacant cell. As you advance, you may develop shortcuts and other tools for this step. However, as you're learning the process, your first step might look something like this:

	A	B	C	D	E	F
1	14	136	1346	2456	1456	1456
2	14	5	2	46	146	3
3	6	12	145	45	3	145
4	15	4	135	356	156	2
5	3	6	45	1	2	456
6	1245	126	1456	3456	456	456

As they advance, most Sudoku players using pencil marks tend to come up with their own systems to help them complete the grids.

Step 2: The next step is to begin your analysis. You want to look for cells where there are only one or two possible numbers that will fit. In this case, 6 can only work for cell B5, thus eliminating the other 6s horizontally, vertically, and in the same sub-grid. This is done by circling the numbers as shown:

	A	B	C	D	E	F
1	14	13⑥	1346	2456	1456	1456
2	14	5	2	46	146	3
3	6	12	145	45	3	145
4	15	4	135	356	156	2
5	3	⑥ **6**	45	1	2	45⑥
6	1245	12⑥	145⑥	3456	456	456

You may also choose to cross out the eliminated pencil marks rather than circling them.

Step 3: Continue your analysis in much the same way, looking for cells where there are only one or two possible numbers that will fit. Once you properly place a number, be sure to eliminate it as a possibility in that row horizontally and vertically, as well as in the same sub-grid, by circling the number in those locations.

Continuing with the analysis, your next moves might be, but are not necessary restricted to, the following:

- B5: The only number possible for that cell is 6.
- C1: This is the only place for the 6 to go in that sub-grid.
- D1: This is the only place for the 2 to go in that sub-grid.
- B1: This is the only place for the 3 to go in that sub-grid.
- C4: This is the only place for the 3 to go in that sub-grid.
- B3: This is the only place for the 2 to go in that sub-grid.
- B6: This is the only place for the 1 to go in that column.
- A6: This is the only place for the 2 to go in that sub-grid.
- D6: This is the only place for the 3 to go in that sub-grid.
- F6: This is the only place for the 6 to go in that column.
- A4: This is the only place for the 5 to go in that column.
- C3: This is the only number remaining for that sub-grid is 1.
- E4: This is the only place for the 1 to go in that sub-grid.
- D4: This is the only place for the 6 to go in that sub-grid.
- D3: This is the only place for the 5 to go in that column.
- D2: 4 is the last remaining number in that column.
- F3: 4 is the last remaining number in that sub-grid.
- F5: 5 is the last remaining number in that row.
- C5: This is the only place for the 4 to go in that row.
- C6: 5 is the last remaining number in that column.
- E6: 4 is the last remaining number in that row.
- E1: 5 is the last remaining number for that cell.
- E2: 6 is the last remaining number for that cell.
- A2: 1 is the last remaining number in that row.
- A1: 4 is the last remaining number in that column.
- F1: 1 is the last remaining number in the Sudoku puzzle.

Of course, the more Sudoku puzzles that your students solve, the more confident, skillful, and adept they become at finding solutions. However, the primary purpose of this book is for all users to master algebraic skills. The Sudoku puzzles are a fun and exciting way for students to check their answers, evaluate their assignments, and assess their progress. Sudoku adds an additional element to the learning process that involves logic, reasoning, and fun, which is intended to encourage students to work with a greater sense of purpose, enthusiasm, eagerness, excitement, and frequency.

Name _____ Date _____ Puzzle #1

GRAPHING METHOD OF SOLVING A SYSTEM OF EQUATIONS

A system of equations consists of two or more equations on a graph. The point of intersection is the solution to the system. Generally, there are three methods of solving a system of equations: (1) graphing method; (2) substitution method; and (3) addition/subtraction. The graphing method involves graphing both equations, and then determining the point of intersection.

Directions: Solve the two systems using the graphing method. Place the coordinates of the solutions in the corresponding cells of the Sudoku grid, and then solve the puzzle.

(B3, A3)

$y = x - 1$

$y = \frac{1}{3}x + 1$

(C2, D2)

$-2x + 2y = -12$

$y + 4 = \frac{1}{2}x$

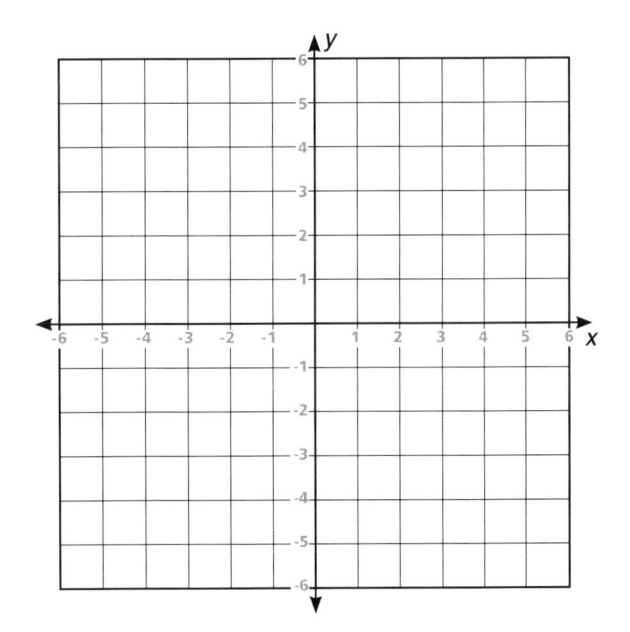

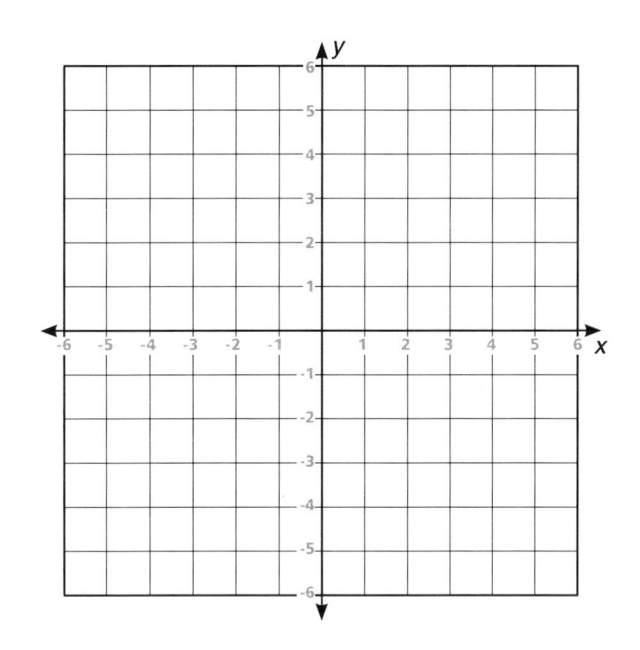

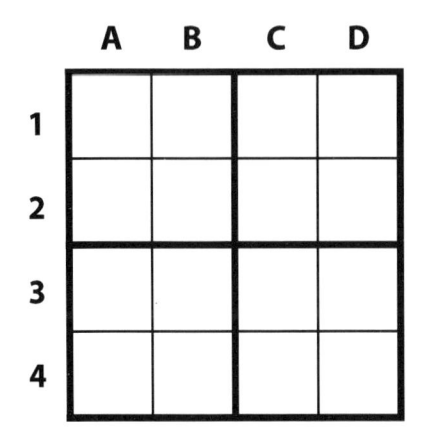

	A	B	C	D
1				
2				
3				
4				

Systems of Equations

Name _____ Date _____ Puzzle #2

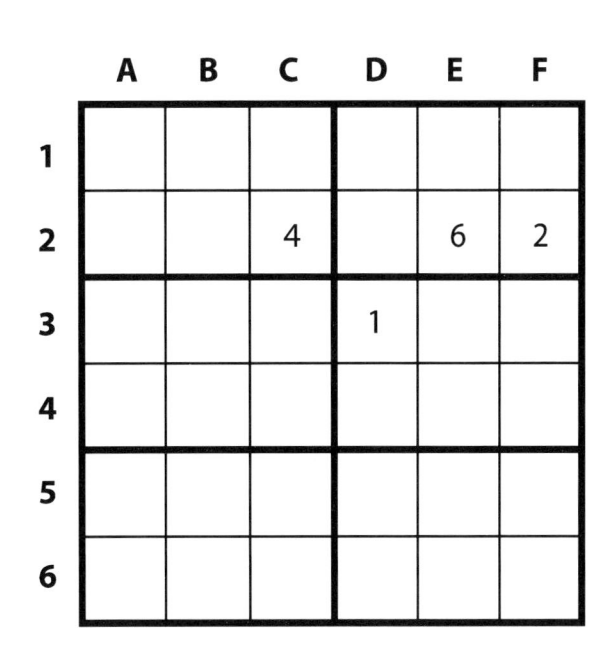

ADDITION/SUBTRACTION METHOD

The addition/subtraction method involves canceling out one of the variables by adding or subtracting the equations. Once one variable is eliminated, it is possible to solve for the other variable. Finally, the value of the known variable is substituted into either equation to solve for the second term, giving you the point of intersection.

Example:

$2x + y = 1$
$\underline{x - y = -4}$
$3x = -3$
$x = -1$

Then substitute -1 for x in either equation:

$2(-1) + y = 1$
$-2 + y = 1$
$y = 3$
Solution: (-1, 3)

Directions: Solve the following systems using the addition/subtraction method. Place the coordinates of the solutions in the corresponding cells of the Sudoku gird, and then solve puzzle. Show all steps!

(A3, A5)
$2x + y = 2$
$x - y = 10$

(B5, C4)
$2x + 2y = 6$
$2x - y = -12$

(F4, D5)
$2x + 4y = -6$
$-3x - 4y = 1$

	A	B	C	D	E	F
1						
2			4		6	2
3				1		
4						
5						
6						

Systems of Equations

11

Name _____   Date _____   Puzzle #3

MORE ADDITION/SUBTRACTION METHOD

At first glance, it might not be obvious that the easiest way to solve the system below is to use the addition/subtraction method. However, it is possible to multiply one of the equations by a number that would allow you to use the method to cancel out a variable.

In this problem, you could either multiply the first equation by 2 or the second equation by -4. Just remember to multiply on both sides of the equation.

$(4x - 3y = 10) \cdot 2 \rightarrow$	$8x - 6y = 20$	OR	$4x - 3y = 10$	$4x - 3y = 10$
$x + 6y = -11$	$\underline{x + 6y = -11}$		$(x + 6y = -11) \cdot -4 \rightarrow$	$\underline{-4x - 24y = 44}$
	$9x = 9$			$-27y = 54$

Directions: Solve the following systems using the addition/subtraction method. Place the coordinates of the solutions in the corresponding cells of the Sudoku gird, and then solve the puzzle. Show all steps!

(A1, A4)
$3x - y = 15$
$x + 2y = -9$

(B1, C3)
$4x + 2y = -6$
$-2x + 4y = -22$

(D4, E6)
$5x + 3y = -7$
$2x - 2y = 10$

	A	B	C	D	E	F
1				2		
2						
3						2
4						
5						
6			2			1

12   *Systems of Equations*

Name _____   Date _____   Puzzle #4

SUBSTITUTION METHOD OF SOLVING SYSTEMS OF EQUATIONS

The substitution method involves solving for one of the variables in one of the equations; substituting that value for the variable into the other equation; and then solving for the other variable. Once one variable is known, the other variable is solved for using either equation.

Example: $2x + 3y = 13$ and $x - y = -1$

It might be easier to solve for x in the second equation, giving us $x = y - 1$. Substitute $y - 1$ for x in the first equation, and then solve. This gives us:

$2(y - 1) + 3y = 13$
$2y - 2 + 3y = 13$
$5y - 2 = 13$
$5y = 15$
$y = 3$

Now substitute 3 for y in either equation.

$x - 3 = -1$
$x = 2$

Solution: (2, 3)

Directions: Solve the following systems using the substitution method. Place the coordinates of the solutions in the corresponding cells of the Sudoku gird, and then solve puzzle. Show all steps!

(C1, A2)
$3y - x = 4$
$x = y + 2$

(C3, E1)
$-2x - y = 8$
$4y + 2 = -3x$

(F5, F6)
$2x = 4y + 16$
$-x - y = 1$

	A	B	C	D	E	F
1	1					
2					5	
3	2					
4				6		5
5		4				
6		1		4		

Systems of Equations

13

Name _____   Date _____   Puzzle #5

SELECTING THE MOST APPROPRIATE METHOD FOR SOLVING A SYSTEM OF EQUATIONS

Directions: For the following systems, determine the most appropriate method for solving the system (graphing, addition/subtraction, or substitution). Then solve each system. Place the coordinates of the solutions in the corresponding cells of the Sudoku grid, and then solve the puzzle. (Show all steps!)

(A4, A5)

$2x - 3y = 6$

$x + 3y = 12$

(B6, D2)

$y = \dfrac{3}{4}x - 2$

$y = -\dfrac{1}{2}x + 3$

(F2, E2)

$2x - 3y = 6$

$x = 3y$

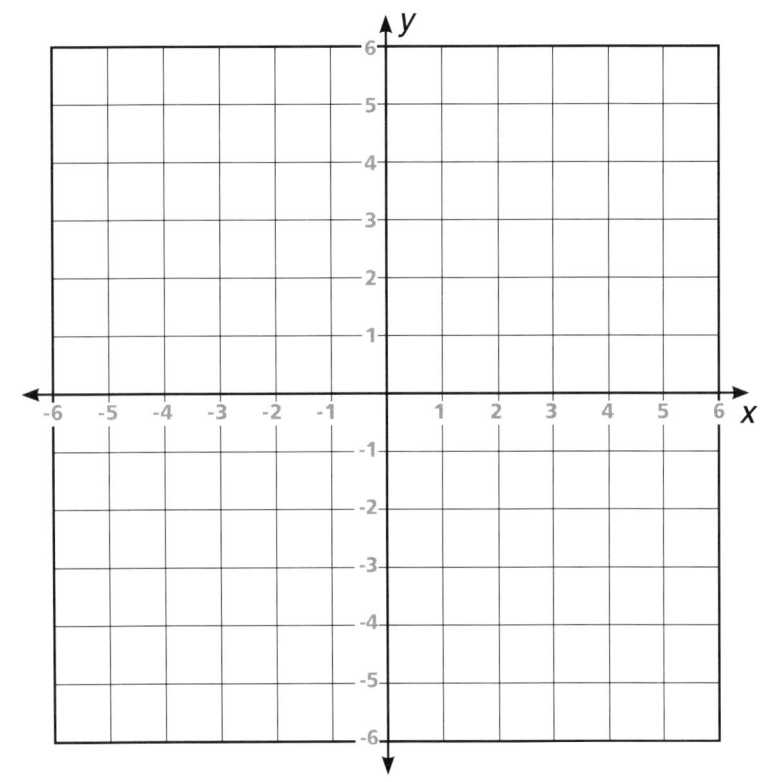

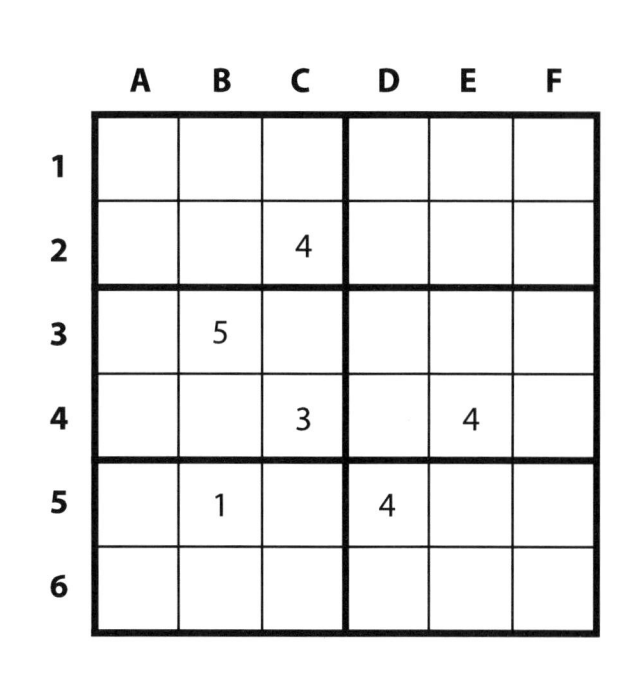

	A	B	C	D	E	F
1						
2			4			
3		5				
4			3		4	
5		1		4		
6						

14   *Systems of Equations*

Name _____ Date _____ Puzzle #6

APPLICATION OF SYSTEMS OF EQUATIONS

Directions: Solve the following word problems using a method of solving a system of equations (graphing, addition/subtraction, or substitution). Place the coordinates of the solutions in the corresponding cells of the Sudoku grid, and then solve the puzzle. Show all work!

(B1, D2) Maxine often babysits for her young cousins, Kala and Olivia. Olivia's age is equal to three-fourths Kala's age minus two years. Olivia's age plus half of Kala's age is equal to 3. How old are Kala and Olivia?

(A2, C2) The school's football team won its first game of the season, scoring 33 points with touchdowns and field goals. One touchdown equals 7 points, and one field goal equals 3 points. If the team scored one more field goal than it did touchdowns, how many touchdowns and field goals did the team score?

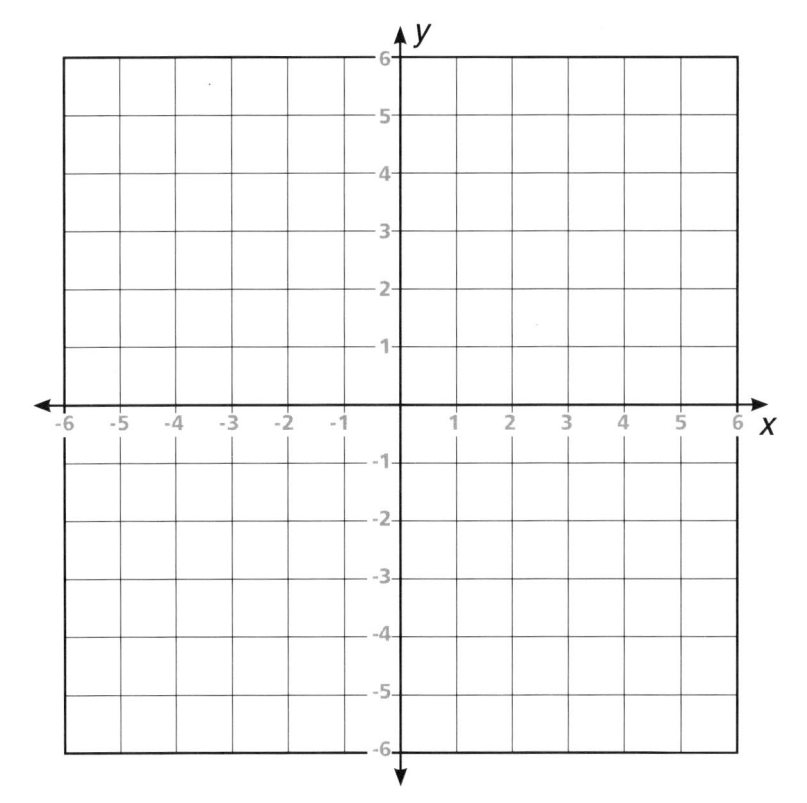

	A	B	C	D
1				
2				
3	2			4
4			3	

Systems of Equations

15

Name _____ Date _____ Puzzle #7

MORE APPLICATION OF SYSTEMS OF EQUATIONS

Directions: Solve the following word problems using a method of solving a system of equations (graphing, addition/subtraction, or substitution). Place the coordinates of the solutions in the corresponding cells of the Sudoku grid, and then solve the puzzle. Show all work!

(A3, E1) Monique's weekly allowance is twice as much as Keyshawn's. Twice Monique's allowance minus three times Keyshawn's allowance equals $2. How much are their allowances?

(B4, D1) Twice Roger's age minus Andy's age is five years. Three times Roger's age plus Andy's age is 20 years. How old are Roger and Andy?

(F1, C2) Tony, St. Mary's star player, scored 22 points in the district championship game. He scored several 3-point and 2-point baskets. If his number of 3-point baskets was two more than twice his number of 2-point baskets, how many 3-point and 2-point baskets did Tony score?

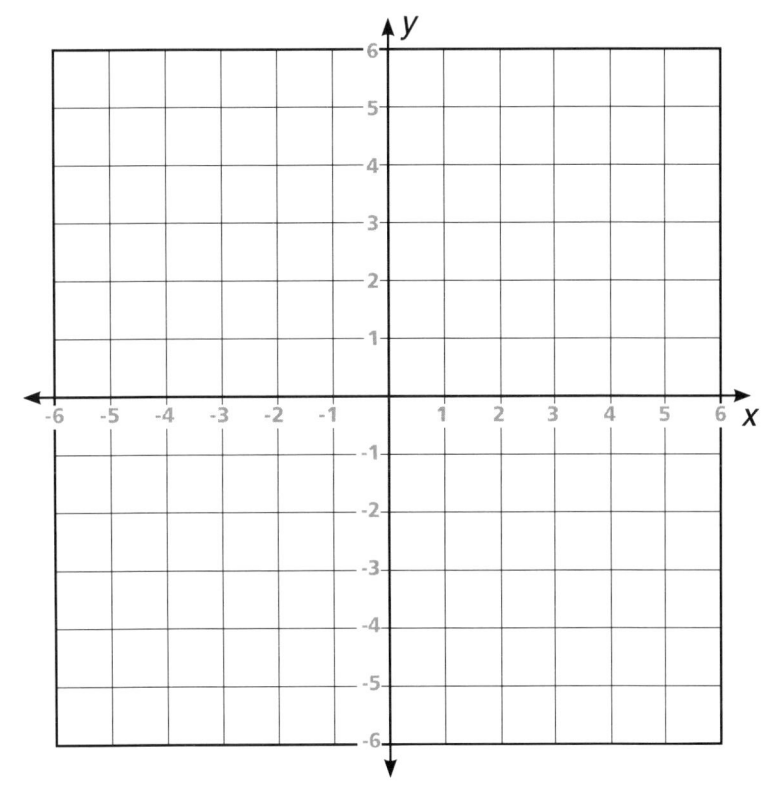

	A	B	C	D	E	F
1		4				
2						
3					3	
4						1
5						
6	5				4	

16   Systems of Equations

Name _____ Date _____ Puzzle #8

SYSTEMS OF EQUATIONS WITH THREE VARIABLES

Directions: Solve this system with three variables using the addition/subtraction method and/or the substitution method. Then use the values for x, y, and z to fill in the corresponding cells of the Sudoku grid. Then solve the puzzle. Show all steps!

$x + y = z$
$x - y = -2z$
$\quad x = z - 3$

	A	B	C	D	E	F
1		y	z			x
2		x		y		z
3	x			z	y	
4	z		y	x		
5	y	z			x	
6			x		y	

Systems of Equations

17

Name _____ Date _____ Puzzle #9

INTRODUCTION TO POLYNOMIALS AND ADDITION

A **polynomial** is one or more groups of algebraic terms separated by plus or minus signs. Certain types of polynomials include:

A *monomial* – one group of terms (e.g. $2x^3y$)
A *binomial* – two groups of terms (e.g. $2xy + 4$)
A *trinomial* – three groups of terms (e.g. $3x^2 + 4x - 7xy^2$)

Directions: Use your knowledge of collecting like terms to add polynomials horizontally or vertically. Find your answer in the box on the right. Place the numbers in the appropriate cells of the Sudoku grid, and then solve the puzzle.

(B1) $(8x^2 - 2y) + (x^2 - 3y)$

(A1) $(7x^2 - xy + y^2) + (3x^2 + 3xy - 6y^2)$

(C3) $(11x^2 + 3xy - 6y^2) + (-x^2 - xy + 2y^2)$

(E6) $(5x^2 + 2xy - y^2) + (5x^2 - xy - y^2)$

(F6) $(9x^2y + x^2 - 4) + (10x^2y^2 - 11x^2y - x^2)$

(D4) $(-x^2 - 6y) + (10x^2 + 2y)$

(E3) $(3x^2 - xy) + (4y^2 - y)$

(1) $9x^2 - 4y$
(3) $10x^2y^2 - 2x^2y - 4$
(3) $10x^2 + 2xy - 4y^2$
(3) $9x^2 - 5y$
(4) $10x^2 + 2xy - 5y^2$
(5) $3x^2 + 4y^2 - xy - y$
(6) $10x^2 + xy - 2y^2$

	A	B	C	D	E	F
1				2		
2						
3						
4		2				
5						
6			2			

18   Polynomials

Name _____ Date _____ Puzzle #10

SUBTRACTING POLYNOMIALS

When subtracting polynomials, reverse all of the signs in the polynomial being subtracted. Then add the polynomials.

For example:

$(8x^2 - 2x) - (6x^2 + 4x) = (8x^2 - 2x) + (-6x^2 - 4x) = 2x^2 - 6x$

Directions: Subtract the polynomials. Find your answer in the box on the right. Place the numbers in the appropriate cells of the Sudoku grid, and then solve the puzzle.

(A6) $(3x^2 - 2x) - (2x^2 + 3x) =$

(B6) $(-6x^2 + x) - (-7x^2 - 6x) =$

(C1) $(8x^2 + 2y - 3) - (-x^2 + 6y + 4) =$

(C4) $(4x^2 - 6y + 5) - (-5x^2 - 7y - 2) =$

(D3) $(-xy - y + x) - (2x + 2y - 4xy) =$

(E4) $(-2x + 2xy + y) - (-x - 2y - xy) =$

(F1) $(2x^2 - 3xy + y^2) - (2x^2 + 3xy + y^2) =$

(2) $-x + 3xy + 3y$
(2) $9x^2 - 4y - 7$
(3) $-6xy$
(3) $-6xy$
(4) $-x - 3y + 3xy$
(4) $x^2 - 5x$
(5) $x^2 + 7x$
(6) $9x^2 + y + 7$

	A	B	C	D	E	F
1					4	
2						
3		1				
4						
5						
6				2		

Polynomials

19

Name _____ Date _____ Puzzle #11

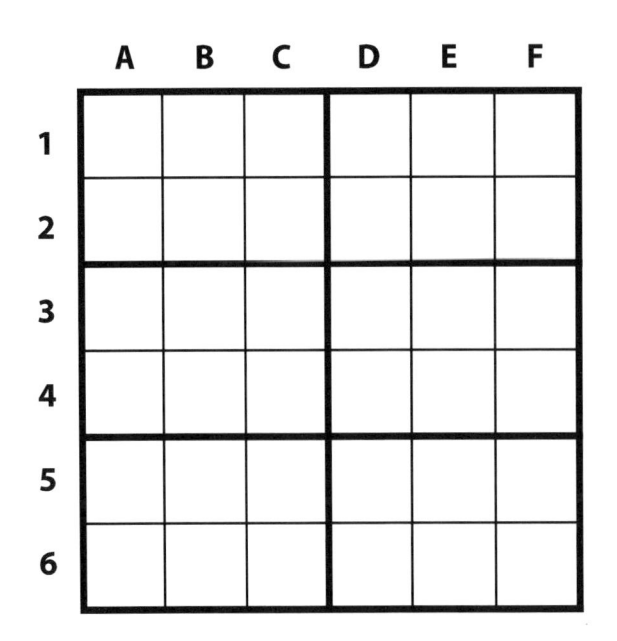

THE FOIL METHOD

The FOIL method is used to multiply two binomials. FOIL is an acronym meaning **first**, **outer**, **inner**, and **last**.

For example: $(3x + 2)(x - 5)$

First: $3x \cdot x = 3x^2$
Outer: $3x \cdot -5 = -15x$
Inner: $2 \cdot x = 2x$
Last: $2 \cdot -5 = -10$

Thus, $3x^2 - 15x + 2x - 10 = 3x^2 - 13x - 10$

Directions: Multiply the binomials using the FOIL method. Find your answer in the box on the right. Place the numbers in the appropriate cells of the Sudoku grid, and then solve the puzzle. Show all steps!

(A2) $(x + 5)(x - 3)$

(A4) $(2x + 1)(x - 3)$

(B3) $(4x + 1)(x - 2)$

(B5) $(x - 4)(4x + 5)$

(C5) $(2x - 1)(2x - 3)$

(D2) $(4x + 1)(x + 2)$

(E2) $(3 - x)(2x - 3)$

(E4) $(x + y)(x + y)$

(F3) $(x - 3)(x - 5)$

(F5) $(x + 2y)(x - 3y)$

(1) $x^2 - 5xy - y^2$
(1) $4x^2 + 9x + 2$
(1) $4x^2 - 11x - 20$
(1) $2x^2 - 5x - 3$
(2) $2x^2 + 5x - 3$
(1) $x^2 + 2xy + y^2$
(3) $4x^2 - 6x - 2$
(1) $-2x^2 + 9x - 9$
(4) $4x^2 - 9x - 2$
(1) $x^2 - xy - 6y^2$
(4) $4x^2 - 7x - 2$
(4) $x^2 + 2x - 15$
(5) $x^2 - 2x + 15$
(1) $x^2 - 8x + 15$
(6) $2x^2 + 9x + 9$
(1) $4x^2 - 8x + 3$

	A	B	C	D	E	F
1						
2						
3						
4						
5						
6						

Polynomials

Name _____ Date _____ Puzzle #12

DIFFERENCE OF SQUARES

If you were to multiply two binomials in the pattern $(a + b)(a - b)$ using the FOIL method, the two middle terms would cancel out. This leaves $a^2 - b^2$, which is a difference of squares.

For example: $(x - 5)(x + 5) = x^2 - 5x + 5x - 25 = x^2 - 25$

Now that you recognize the pattern, the process is possible without going through the FOIL method. Instead, square the two terms and put a minus sign in between them.

Directions: Multiply the binomials. Find your answer in the box on the right. Place the numbers in the appropriate cells of the Sudoku grid, and then solve the puzzle.

(A3) $(x + 2)(x - 2)$	(E6) $(3x + 1)(3x - 1)$	(4) $x^2 - 16$	(1) $y^2 - 144$
(A4) $(x + 4)(x - 4)$	(E8) $(3x - 2)(3x + 2)$	(8) $x^2 - 100$	(5) $4x^2 - 9y^2$
(A6) $(x - 5)(x + 5)$	(F1) $(4 - x)(4 + x)$	(6) $4x^2 - 1$	(2) $x^2 - 4$
(A7) $(x - y)(x + y)$	(F3) $(x^2 + 1)(x^2 - 1)$	(7) $x^2 - 64$	(1) $16x^4 - 1$
(A9) $(x - 8)(x + 8)$	(F7) $(5 - y)(5 + y)$	(4) $y^2 - 4x^2$	(1) $16x^2 - 1$
(B5) $(y - 6)(y + 6)$	(G2) $(4x + 1)(4x - 1)$	(8) $x^4 - 9$	(7) $36 - x^2$
(B9) $(x + 1)(x - 1)$	(G3) $(2x - 3y)(2x + 3y)$	(7) $x^4 - y^4$	(1) $x^2 - y^2$
(C2) $(x + 10)(x - 10)$	(G4) $(x + xy)(x - xy)$	(6) $81 - x^2y^2$	(3) $x^4y^4 - 1$
(C4) $(x - 7)(x + 7)$	(G6) $(6 - x)(6 + x)$	(2) $x^2 - 81$	(8) $y^2 - z^2$
(C6) $(x - 3)(x + 3)$	(G8) $(x^2 + 3)(x^2 - 3)$	(9) $x^2y^2 - 100$	(7) $x^2 - 49$
(C7) $(x + 9)(x - 9)$	(H1) $(x^2 - y^2)(x^2 + y^2)$	(2) $x^2 - 169$	(5) $x^2 - 25y^2$
(C8) $(x - 11)(x + 11)$	(H4) $(4x^2 + 1)(4x^2 - 1)$	(3) $9x^2 - 4$	(3) $x^2 - x^2y^2$
(D3) $(y + 12)(y - 12)$	(I1) $(9 - xy)(9 + xy)$	(1) $y^2 - 36$	(8) $x^2 - 1$
(D7) $(2x + 1)(2x - 1)$	(I3) $(y - z)(y + z)$	(8) $25 - y^2$	(9) $4x^2 - 9$
(D9) $(2x - 3)(2x + 3)$	(I4) $(-x + 5y)(-x - 5y)$	(9) $x^2 - 121$	(7) $x^4 - 1$
(E2) $(y - 2x)(y + 2x)$	(I6) $(xy - 10)(xy + 10)$	(2) $16 - x^2$	(1) $9x^2 - 1$
(E4) $(x + 13)(x - 13)$	(I7) $(x^2y^2 - 1)(x^2y^2 + 1)$	(3) $x^2 - 25$	(6) $x^2 - 9$

```
     A   B   C   D   E   F   G   H   I
  1 [                                   ]
  2 [                                   ]
  3 [                                   ]
  4 [                                   ]
  5 [                                   ]
  6 [                                   ]
  7 [                                   ]
  8 [                                   ]
  9 [                                   ]
```

Polynomials

21

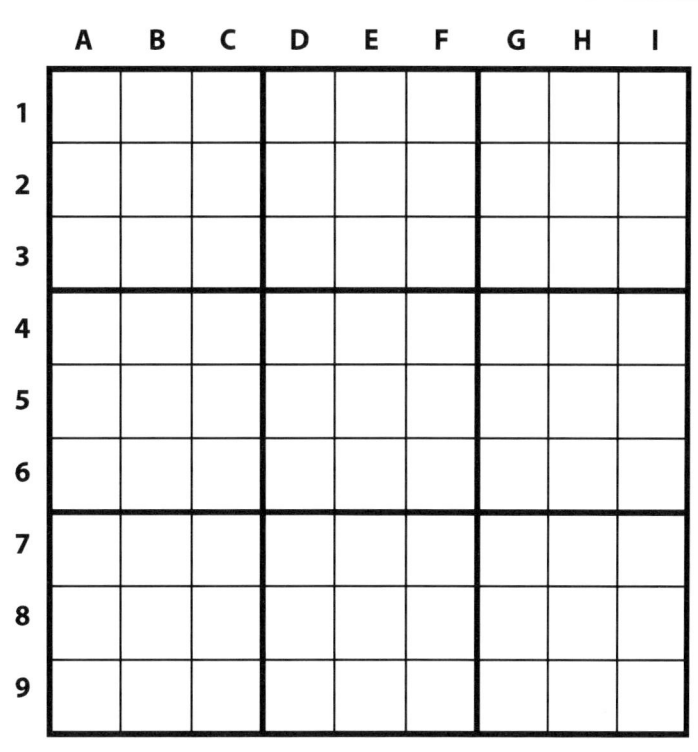

Name _____ Date _____ Puzzle #13

PERFECT-SQUARE TRINOMIALS

When a binomial is squared, the result is a perfect-square trinomial.

For example: $(x+3)^2 = (x+3)(x+3)$
$x^2 + 3x + 3x + 9 = x^2 + 6x + 9$

$x^2 + 6x + 9$ is considered a perfect-square trinomial because the first term (x^2) and the last term (9) are squares, and the middle term is the product of the square roots of the first and last terms doubled ($x \cdot 3 = 3x \cdot 2 = 6x$). Now that you recognize the pattern, the process is possible without using the FOIL method.

Directions: Square the following binomials. Find your answer in the box on the right. Place the numbers in the appropriate cells of the Sudoku grid, and then solve the puzzle.

(A4) $(x+3)^2$

(A6) $(x-5)^2$

(B2) $(2x+1)^2$

(B3) $(x+7)^2$

(C2) $(x-1)^2$

(C4) $(3x-2)^2$

(D3) $(x+y)^2$

(E5) $(5x-2)^2$

(F3) $(x^2+3)^2$

(1) $9x^2 - 12x + 4$
(1) $x^2 - 5x + 25$
(2) $4x^2 + 4x + 1$
(2) $x^2 + 2x - 1$
(2) $x^2 - 10x + 25$
(3) $x^2 + 10x - 25$
(4) $x^4 + 6x^2 + 9$
(4) $x^2 - 2x + 1$
(4) $x^2 + 6x + 9$
(5) $x^2 + 3x + 9$
(5) $25x^2 - 20x + 4$
(5) $x^2 + 2xy + y^2$
(6) $x^2 + xy + y^2$
(6) $x^2 + 14x + 49$

	A	B	C	D	E	F
1						5
2						
3						
4					2	
5				3		
6						

22 Polynomials

Name _____ Date _____ Puzzle #14

MULTIPLYING POLYNOMIALS

Multiplying by binomials or trinomials horizontally is similar to multiplying by a two- or three-digit whole number.

For example: $(2x^2 + 3x - 6) \cdot (x + 3)$

1) Multiply 3 times each term of the trinomial. $3(2x^2 + 3x - 6) = 6x^2 + 9x - 18$

2) Now multiply x times each term of the trinomial. $x(2x^2 + 3x - 6) = 2x^3 + 3x^2 - 6x$
    Be sure to line-up like terms!

3) Add. $2x^3 + 3x^2 - 6x + 6x^2 + 9x - 18$
    $2x^3 + 9x^2 + 3x - 18$

Directions: Multiply the polynomials below. Find your answer in the box on the right. Place the numbers in the appropriate cells of the Sudoku grid, and then solve the puzzle. Show all steps!

(A2) $2x^2 - 3x + 1$
         $x + 4$

(B1) $3x^2 - x + 2$
         $2x - 1$

(1) $6x^3 - 5x^2 + 5x - 2$
(2) $-5x^3 + 14x^2 + 19x - 6$
(2) $3x^3 + y^2 - 2xy - 4x + 4y$
(3) $3x^2 - 4x - 4xy + y^2 + 4y$
(4) $-5x^3 - 8x^2 + 19x - 6$
(4) $2x^3 + 5x^2 - 11x + 4$

(C4) $-x^2 - 2x + 3$
         $5x - 2$

(D3) $3x - y - 4$
         $x - y$

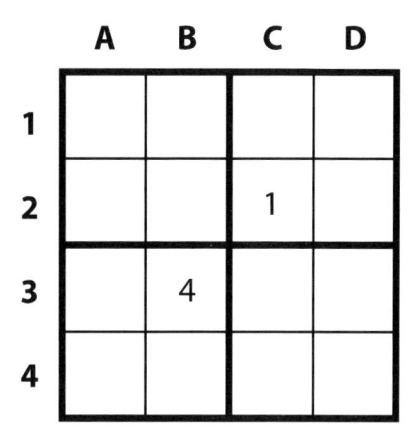

	A	B	C	D
1				
2			1	
3		4		
4				

Polynomials

23

Name _____ Date _____ Puzzle #15

DIVIDING POLYNOMIALS

Dividing by a binomial or trinomial can be done in a way similar to long division.

Follow along with this example:

1) 2x can be divided into $2x^2$, x times.
2) Multiply by x.
3) Subtract.
4) 2x can be divided into -12x, -6 times.
5) Subtract.

$$\begin{array}{r} x - 6 \\ 2x + 1 \overline{\smash{)}\ 2x^2 - 11x - 6} \\ -\underline{(2x^2 + x)} \\ -12x - 6 \\ -\underline{(-12x - 6)} \\ 0 \end{array}$$

Directions: Divide the polynomials below, Find your answer in the box on the right. Place the numbers in the appropriate cells of the Sudoku grid, and then solve the puzzle. Show all steps!

(A1) $3x + 1 \overline{\smash{)}\ 3x^2 - 14x - 5}$

(A2) $2x - 1 \overline{\smash{)}\ 2x^2 + 7x - 4}$

(1) $x + 5$
(1) $x + 3$
(2) $x - 5$
(2) $x - 4$
(3) $x - 2$
(3) $x - 3$
(4) $x + 2$
(4) $x + 4$
(4) $2x^2 - 1$

(A4) $4x - 3 \overline{\smash{)}\ 4x^2 - 11x + 6}$

*(D4) $2x^3 + 1 \overline{\smash{)}\ 2x^4 + 6x^3 + \underline{} + x + 3}$

* *Fill in a place holder for the descending power of the variable, such as $0x^2$.*

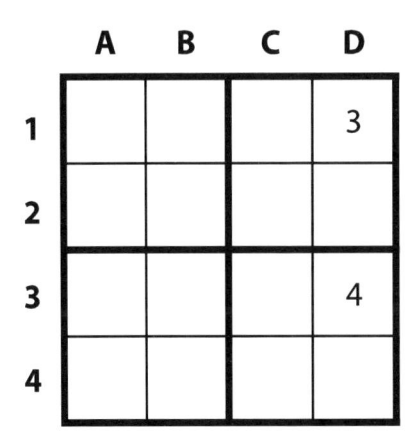

	A	B	C	D
1				3
2				
3				4
4				

24

Polynomials

Name _____ Date _____ Puzzle #16

GREATEST COMMON MONOMIAL FACTOR

The Greatest Common Monomial Factor (GCMF) is the greatest term that divides evenly into a group of terms.

For example, the GCMF of $4x^3$, $6x^4$, and $8x^2$ is $2x^2$.

In order for a variable to be common to a group of terms, it must be represented in each term. If the variable(s) is present in each group of terms, choose the lowest power of the variable as part of your GCMF.

Here are other examples of the GCMF:

A) For $9x^5y^4$, $12x^4y^3$, and $15x^7y^5$ the GCMF is $3x^4y^3$.

B) For $20x^3y$, $15x^2$, and $10y^2$ the GCMF is 5. (x and y are not represented in all terms.)

Directions: Determine the GCMF for each group of terms. Find your answer in the box on the right. Place the numbers in the appropriate cells of the Sudoku grid, and then solve the puzzle.

(A3) $4x^3$, $10x^4$, $12x^2$

(B4) $8x^7$, $16x^4$, $20x^5$

(C3) $9x^4$, $18y^3$, $27x^2y^3$

(D6) $6x^2y^3$, $9x^3z^2$, $12x^4$

(E3) $15x^4y^5$, $10x^3y^7$, $20x^5y^3$

(F2) $25x^3y^4z^2$, $15x^2y^5$, $3y^3z^4$

(1) y^3
(2) $3x^2$
(3) $5x^3y^3$
(4) $2x^2$
(5) 9
(6) $4x^4$

	A	B	C	D	E	F
1			3			
2					2	
3						
4				5		2
5	5	3				
6						

Factoring

25

Name _____ Date _____ Puzzle #17

FACTORING THE GCMF

Factoring out the Greatest Common Monomial Factor (GCMF) is like working the distributive property in reverse. First, determine the GCMF of the polynomial. Put the GCMF outside a group of parentheses. Then place inside the parentheses the number that could be multiplied times the GCMF to give you the original polynomial.

Here are some examples:

A) $(4x^3 - 8x^2 + 6x) = 2x(2x^2 - 4x + 3)$

B) $(-9x^3y^2 + 12x^2y^3 - 15x^4y) = -3x^2y(3xy - 4y^2 + 5x^2)$
(If the first term of the polynomial is negative, factor out a negative in your GCMF.)

Directions: Factor the GCMF. Find your answer in the box on the right. Place the numbers in the appropriate cells of the Sudoku grid, and then solve the puzzle.

(A4) $(4x^5 + 8x^4)$

(B3) $(20x^4 - 15x^3)$

(C1) $(9x^2 - 6y^2 - 12z^3)$

(D4) $(-8x^4y^2 + 10x^3y^5 - 12x^2y^4)$

(E2) $(12x^3y^3 + 18x^4y^2 - 24y^4)$

(F2) $(-8x^2z^3 + 16x^3z^5 - 12xz^6)$

(1) $-2x^2y^2(4x^2 - 5xy^3 + 6y^2)$
(2) $-4xz^3(2x - 4x^2z^2 + 3z^3)$
(3) $4x^4(x + 2)$
(4) $6y^2(2x^3y + 3x^4 - 4y^2)$
(5) $5x^3(4x - 3)$
(6) $3(3x^2 - 2y^2 - 4z^3)$

	A	B	C	D	E	F
1						
2						
3			1			6
4				2		
5	6	1				
6				2		

26

Name _____   Date _____   Puzzle #18

FACTORING A DIFFERENCE OF SQUARES

When factoring a difference of squares it is important to remember that $(a + b)(a - b) = a^2 - b^2$.

For example: $x^2 - 4$
This binomial is a difference of squares and can be factored $(x + 2)(x - 2)$.

Directions: Factor the following binomials. Find your answer in the box on the right. Place the numbers in the appropriate cells of the Sudoku grid, and then solve the puzzle.

(A1) $x^2 - 9$	**(E6)** $x^2 - 81$
(A4) $x^2 - 1$	**(F1)** $x^2 - 121$
(A6) $x^2 - 16$	**(F3)** $36x^2 - 1$
(B3) $x^2 - 25$	**(F6)** $x^2 - y^2$
(B6) $x^2 - 36$	**(F9)** $x^2y^2 - 4$
(B8) $x^2 - 144$	**(G7)** $x^2 - 50$
(C2) $4x^2 - 1$	**(H4)** $25x^2 - 4$
(C3) $x^2 - 49$	**(H7)** $16 - y^2$
(D4) $9x^2 - 4$	**(I1)** $1 - 9x^2$
(D7) $x^2 - 64$	**(I4)** $x^2 - 225$
(D9) $x^2 - 100$	**(I6)** $16x^2y^2 - z^2$
(E4) $1 - x^2$	**(I9)** $x^2 - 169$

(7) $(x + 5)(x - 5)$	**(3)** $(x - 6)(x + 6)$
(8) $(x + 9)(x - 9)$	**(5)** $(1 + x)(1 - x)$
(4) $(5x + 2)(5x - 2)$	**(1)** $(x - 15)(x + 15)$
(6) $(1 - 3x)(1 + 3x)$	**(2)** $(4xy + z)(4xy - z)$
(7) $(x + 13)(x - 13)$	**(8)** $(4 - y)(4 + y)$
(2) $(x + 1)(x - 1)$	**(4)** $(x + 3)(x - 3)$
(1) Prime	**(8)** $(6x - 1)(6x + 1)$
(8) $(x - 10)(x + 10)$	**(6)** $(x + y)(x - y)$
(2) $(xy + 2)(xy - 2)$	**(7)** $(x + 8)(x - 8)$
(9) $(3x - 2)(3x + 2)$	**(1)** $(x - 11)(x + 11)$
(3) $(x + 7)(x - 7)$	**(8)** $(2x + 1)(2x - 1)$
(5) $(x - 4)(x + 4)$	**(1)** $(x + 12)(x - 12)$

	A	B	C	D	E	F	G	H	I
1		2		3					
2								1	
3									
4									
5					4				
6									
7									
8							4		
9	9							3	

Factoring

27

Name _____ Date _____ Puzzle #19

FACTORING PERFECT-SQUARE TRINOMIALS

In a perfect-square trinomial, the first and last terms are squares, and the middle term is the product of the square roots of the first and last terms doubled. A factored perfect-square trinomial is a binomial squared.

For example, $4x^2 + 12x + 9$ is a perfect square trinomial.
Factored, it is $(2x + 3)^2$.

Directions: Factor the perfect square trinomials. (Be careful, one is prime or not factorable). Find your answer in the box on the right. Place the numbers in the appropriate cells of the Sudoku grid, and then solve the puzzle.

(A3) $x^2 - 8x + 16$	(D4) $4x^2 + 20x + 25$
(A5) $x^2 + 10x + 25$	(D6) $9x^2 + 30x + 25$
(B4) $x^2 - 2x + 1$	(E2) $25x^2 + 10x + 4$
(B5) $x^2 + 18x + 81$	(E3) $16x^2 - 8x + 1$
(C1) $x^2 - 14x + 49$	(F2) $9x^2 + 24x + 16$
(C3) $x^2 + 4x + 4$	(F4) $4x^2 - 4xy + y^2$

(1) $(x + 5)^2$	(2) $(x + 9)^2$
(1) $(x + 2)^2$	(1) $(2x + 5)^2$
(4) $(3x + 4)^2$	(6) $(2x - y)^2$
(2) $(4x - 1)^2$	(6) Prime
(6) $(3x + 5)^2$	(5) $(x - 4)^2$
(3) $(x - 1)^2$	(2) $(x - 7)^2$

28

Factoring

Name _____   Date _____   Puzzle #20

SIMPLE TRINOMIALS

Simple trinomials could have easily been called the "fun" trinomials, because they are like solving a puzzle – fun!

Let's use $x^2 + 5x + 6$, a simple trinomial. To factor it, find two numbers that you can multiply together to get 6, and add together to get 5 (the coefficient of x). Those two numbers are 2 and 3, because 2 times 3 equals 6, and 2 plus 3 equals 5. So the factored trinomial would be: $(x + 2)(x + 3)$.

(Note: In the simple trinomial pattern, there is no coefficient for x^2.)

Directions: Factor the simple trinomials. Find your answer in the box on the right. Place the numbers in the appropriate cells of the Sudoku grid, and then solve the puzzle.

(A1) $x^2 + 6x + 8$	**(E8)** $x^2 - 2x + 15$
(A4) $x^2 + 2x - 15$	**(F7)** $x^2 - 12x + 11$
(A9) $x^2 - 5x + 6$	**(F9)** $x^2 - 9x - 36$
(B2) $x^2 + 5x + 4$	**(G5)** $x^2 - 11x + 28$
(B6) $x^2 - 3x - 28$	**(G7)** $x^2 - 9x + 14$
(C3) $x^2 - 5x - 14$	**(H2)** $x^2 + 5x - 24$
(C5) $x^2 - 11x + 24$	**(H6)** $x^2 + 11x - 26$
(D1) $x^2 + 22x - 75$	**(I1)** $x^2 + 3x - 54$
(D9) $x^2 - x - 30$	**(I4)** $x^2 - 20x + 36$
(E2) $x^2 - 15x + 56$	**(I7)** $x^2 + 3x - 28$

(5) $(x - 3)(x - 2)$	**(3)** $(x + 2)(x + 4)$
(6) $(x + 13)(x - 2)$	**(2)** $(x - 2)(x - 18)$
(3) $(x - 6)(x + 5)$	**(4)** $(x + 7)(x - 4)$
(8) $(x + 9)(x - 6)$	**(3)** $(x - 2)(x - 7)$
(4) $(x - 7)(x + 4)$	**(6)** $(x - 3)(x - 8)$
(3) $(x + 8)(x - 3)$	**(9)** $(x - 1)(x - 11)$
(1) $(x - 7)(x - 4)$	**(6)** $(x + 3)(x - 12)$
(4) $(x - 7)(x + 2)$	**(2)** $(x - 7)(x - 8)$
(5) Prime	**(4)** $(x - 3)(x + 25)$
(1) $(x + 4)(x + 1)$	**(7)** $(x + 5)(x - 3)$

	A	B	C	D	E	F	G	H	I
1		7				1			
2									
3	6			8					
4		5			8			4	
5									
6	9				1				5
7									
8		3						9	
9								8	1

Factoring

29

Name _____ Date _____ Puzzle #21

FACTOR BY GROUPING

Factoring by grouping is used to factor certain polynomials in four terms.

Using the polynomial $2x^2 + 2x + 3x + 3$, follow along to see how it is done:

1) Group the terms into two pairs. $(2x^2 + 2x) + (3x + 3)$

2) Remove the GCMF from each group. $2x(x + 1) + 3(x + 1)$

3) Notice that the two quantities in parentheses are now identical! We can now factor out the common factor $(x + 1)$ using the distributive property, giving us: $(2x + 3)(x + 1)$

Directions: Factor by grouping. Find your answer in the box on the right. Place the numbers in the appropriate cells of the Sudoku grid, and then solve the puzzle. Show all steps!

(A4) $5x^2 + 15x + 2x + 6$

(B3) $2x^2 - 2x - 3x + 3$

(C2) $6x^2 + 9x + 4x + 6$

(D4) $6x^3 - 9x^2 + 2xy - 3y$

(E1) $2x^2 + 3x - 2xy - 3y$

(E4) $2x^3 - 2x^2 - 3x + 3$

(F1) $5x^2z - 2yz - 15x^2 + 6y$

(F3) $-2xy + 10 - xy^2 + 5y$

(1) $(x - y)(2x + 3)$
(2) $(2x - 3)(x - 1)$
(3) $(5x + 2)(x + 3)$
(4) $(z - 3)(5x^2 - 2y)$
(4) $(2x^2 - 3)(x - 1)$
(5) $(3x + 2)(2x + 3)$
(5) $(-2 - y)(xy - 5)$
(6) $(3x^2 + y)(2x - 3)$

	A	B	C	D	E	F
1						
2						
3			6			
4						
5				4		
6	5	6				

30   Factoring

Name _____   Date _____   Puzzle #22

GENERAL TRINOMIALS

General trinomials are in the form $ax^2 + bx + c$. For example, $6x^2 + 11x + 4$ is a general trinomial. Perhaps the simplest method of factoring general trinomials is to factor using the grouping method. Here is how it's done:

1) Find the grouping number by multiplying $a \cdot c$. In the case of $6x^2 + 11x + 4$, the grouping number is 24 (6 • 4).

2) Find two numbers that we can multiply together to get 24 and add together to get b (which is 11 in this problem). Those numbers are 3 and 8, because 3 • 8 = 24 and 3 + 8 = 11.

3) Rewrite the trinomial in four terms, substituting $3x + 8x$ for $11x$. This leaves $6x^2 + 3x + 8x + 4$.

4) Proceed to factor by group: $(6x^2 + 3x) + (8x + 4) = 3x(2x + 1) + 4(2x + 1)$
    Answer: $(2x + 1)(3x + 4)$

Directions: Factor the general trinomials. Find your answer in the box on the right. Place the numbers in the appropriate cells of the Sudoku grid, and then solve the puzzle. Show all steps!

(B2) $2x^2 + 9x + 10$

(1) $(4x - 3)(x + 2)$
(2) $(2x - 5)(3x - 4)$
(3) $(3x - 4)(2x + 5)$
(3) $(3x - 2)(x + 4)$
(4) $(2x + 5)(x + 2)$

(B3) $4x^2 + 5x - 6$

(B4) $3x^2 + 10x - 8$

(C3) $6x^2 - 23x + 20$

	A	B	C	D
1			3	
2				
3				
4				

Factoring

Name _____ Date _____ Puzzle #23

FACTORING COMPLETELY

Factoring completely involves considering all possible factoring patterns, starting with the Greatest Common Monomial Factor (GCMF). After the GCMF, look to factor the following patterns in order:

A) Difference of Squares: $3x^2 - 12 = 3(x^2 - 4) = 3(x+2)(x-2)$

B) Perfect-Square Trinomials: $4x^3 - 12x^2 + 9x = x(4x^2 - 12x + 9) = x(2x-3)^2$

C) Simple Trinomials: $2x^2 - 14x - 36 = 2(x^2 - 7x - 18) = 2(x-9)(x+2)$

D) General Trinomials: $4x^3y - 11x^2y - 3xy = xy(4x^2 - 11x - 3) = xy(4x^2 - 12x + x - 3) =$
$xy[4x(x-3) + 1(x-3)] = xy(4x+1)(x-3)$

Directions: Factor completely. Find your answer in the box on the right. Place the numbers in the appropriate cells of the Sudoku grid, and then solve the puzzle. Show all steps!

(A4) $6x^2 - 54$

(B3) $2x^3 - 6x^2 - 56x$

(C1) $18x^2 + 60x + 50$

(C3) $3x^2y + 14xy - 5y$

(D6) $x^3y - 7x^2y + 6xy$

(E2) $-12x^2 + 36x - 27$

(F2) $8x^4y^2 + 2x^3y^2 - 10x^2y^2$

(F3) $3x^4 - 3$

(1) $2x(x-7)(x+4)$
(2) $2(3x+5)^2$
(2) $3(x+1)(x-1)(x^2+1)$
(3) $-3(2x-3)^2$
(4) $6(x+3)(x-3)$
(5) $y(3x-1)(x+5)$
(6) $2x^2y^2(4x+5)(x-1)$
(6) $xy(x-6)(x-1)$

	A	B	C	D	E	F
1						
2						
3						
4				5	6	
5	2	5				
6						

32

Factoring

Name _____ Date _____ Puzzle #24

A RADICAL INTRODUCTION

As you know, the **square root** of a number is a number which, when multiplied by itself, produces the given number. The symbol for a square root is the radical ($\sqrt{\ }$).

Examples:

$\sqrt{49} = 7$ $\qquad$ $\sqrt{x^6} = x^3$ $\qquad$ $\sqrt{25x^4} = 5x^2$

Directions: Find the square roots of the radicals below. Find your answer in the box on the right. Place the numbers in the appropriate cells of the Sudoku grid, and then solve the puzzle.

(A2) $\sqrt{64}$

(B2) $\sqrt{x^6}$

(B3) $\sqrt{x^2 y^4}$

(C4) $\sqrt{64x^8 y^{10}}$

(D1) $\sqrt{25y^4 z^{12}}$

(E4) $\sqrt{169x^{14} y^2 z^8}$

(F3) $\sqrt{\frac{1}{4}}$

(F5) $\sqrt{0.36x^4}$

(1) x^3
(2) 8
(3) $13x^7 yz^4$
(4) $8x^4 y^5$
(4) $0.6x^2$
(5) xy^2
(5) $5y^2 z^6$
(6) $\frac{1}{2}$

	A	B	C	D	E	F
1						
2						
3				4		
4	1					
5					5	
6			1			

Radicals

Name _____ Date _____ Puzzle #25

SIMPLIFYING RADICALS

When the **radicand** (the term under radical) is not a perfect square, break it down into the product of two radicals: one for the GCF (Greatest Common Factor) that is a perfect square, and the other for the remaining factor. Then simplify.

Examples:

A) $\sqrt{x^5} = \sqrt{x^4}\sqrt{x} = x^2\sqrt{x}$

B) $\sqrt{8x^3} = \sqrt{4x^2}\sqrt{2x} = 2x\sqrt{2x}$

Directions: Simplify the radicals. Find your answer in the box on the right. Place the numbers in the appropriate cells of the Sudoku grid, and then solve the puzzle. Show all steps!

(A2) $\sqrt{x^7}$

(B2) $\sqrt{50}$

(B3) $\sqrt{4x^5}$

(C4) $\sqrt{x^7 y^9}$

(D1) $\sqrt{27y^4 x^3}$

(D3) $\sqrt{32xy^7}$

(E4) $\sqrt{200x^4 y^7}$

(F3) $\sqrt{98x^9 y^{11} z}$

(1) $2x^2\sqrt{x}$
(1) $3y^2 x\sqrt{3x}$
(2) $5\sqrt{2}$
(3) $7x^4 y^5 \sqrt{2xyz}$
(4) $10x^2 y^3 \sqrt{2y}$
(5) $x^3 y^4 \sqrt{xy}$
(5) $4y^3 \sqrt{2xy}$
(6) $x^3 \sqrt{x}$

	A	B	C	D	E	F
1						
2						
3						
4	2					
5					1	5
6			2			

Radicals

Name _____ Date _____ Puzzle #26

ADDING AND SUBTRACTING RADICAL EXPRESSIONS

Adding and subtracting expressions containing radicals is similar to collecting to like terms, provided the terms beneath the radical (known as the radicand) are the same.

Examples:

(A) $4\sqrt{3} + 5\sqrt{3} = 9\sqrt{3}$

(B) $3\sqrt{x} + \sqrt{y} - \sqrt{x} = 2\sqrt{x} + \sqrt{y}$

Directions: Simplify the radical expressions. Find your answer in the box on the right. Place the numbers in the appropriate cells of the Sudoku grid, and then solve the puzzle.

(A4) $3\sqrt{5} - \sqrt{5} =$

(A5) $3\sqrt{2} + 2\sqrt{2} + \sqrt{2} =$

(B3) $2\sqrt{3} + 2\sqrt{7} =$

(C3) $6\sqrt{x} - \sqrt{x} =$

(D4) $\sqrt{3} - 4\sqrt{3} + 2\sqrt{3} =$

(E2) $-\sqrt{x} - 2\sqrt{xy} + \sqrt{xy} + 3\sqrt{x} =$

(F2) $\sqrt{11} + \sqrt{7} + \sqrt{11} - 3\sqrt{7} =$

(F3) $3\sqrt{13} - 4\sqrt{11} - \sqrt{11} - 5\sqrt{13} + 2\sqrt{13} - 5\sqrt{11} =$

(1) $2\sqrt{11} - 2\sqrt{7}$
(2) $-10\sqrt{11}$
(2) $6\sqrt{2}$
(3) $2\sqrt{x} - \sqrt{xy}$
(4) $2\sqrt{3} + 2\sqrt{7}$
(5) $2\sqrt{5}$
(6) $5\sqrt{x}$
(6) $-\sqrt{3}$

	A	B	C	D	E	F
1			2			
2						
3						
4				1		
5		6				
6				1		

Radicals 35

Name _____ Date _____ Puzzle #27

MULTIPLYING RADICALS

$\sqrt{a} \cdot \sqrt{b} = \sqrt{ab}$: This rule makes multiplying radicals easy.

Examples:

$\sqrt{3} \cdot \sqrt{2} = \sqrt{6}$ $3\sqrt{2} \cdot 2\sqrt{8} = 6\sqrt{16} = 6 \cdot 4 = 24$

Directions: Multiply these radical expressions, simplifying your final answers if needed. Find your answer in the box on the right. Place the numbers in the appropriate cells of the Sudoku grid, and then solve the puzzle.

(A3) $\sqrt{3} \cdot \sqrt{12}$

(A6) $\sqrt{7} \cdot \sqrt{3}$

(B4) $3\sqrt{5} \cdot 4\sqrt{7}$

(C3) $\sqrt{3} \cdot \sqrt{6}$

(D5) $2\sqrt{2} \cdot 5\sqrt{6}$

(E3) $4\sqrt{6}\,(2\sqrt{24} - 3\sqrt{6})$

(F1) $(\sqrt{3} + 2)(\sqrt{3} - 2)$ (Hint: Use the FOIL method)

(F4) $(\sqrt{3} + 4)(1 - \sqrt{12})$

(1) 6
(2) $-7\sqrt{3} - 2$
(3) 24
(4) $12\sqrt{35}$
(4) $20\sqrt{3}$
(5) $\sqrt{21}$
(6) $3\sqrt{2}$
(6) -1

	A	B	C	D	E	F
1					4	
2			1			
3						
4				6		
5						
6		1				

36   Radicals

Name _____   Date _____   Puzzle #28

RATIONALIZING THE DENOMINATOR

When a radical remains in the denominator, it is considered an **irrational number**.

For example, $\dfrac{2}{\sqrt{5}}$ is an irrational number.

In order to rationalize the number, it is important to get rid of all radicals that are in the denominator. To do this, multiply by the square root of an expression that will give a perfect square under the radical in the denominator:

$$\dfrac{2}{\sqrt{5}} \cdot \dfrac{\sqrt{5}}{\sqrt{5}} = \dfrac{2\sqrt{5}}{5}$$

Directions: Rationalize the following denominators. Find your answer in the box on the right. Place the numbers in the appropriate cells of the Sudoku grid, and then solve the puzzle. Show all work!

(A2) $\dfrac{3}{\sqrt{5}}$

(A3) $\dfrac{\sqrt{8}}{2\sqrt{2}}$

(B3) $\dfrac{3\sqrt{18}}{\sqrt{2}}$

(D3) $\dfrac{2\sqrt{3}}{\sqrt{3}+1}$ (Hint: Think difference of squares)

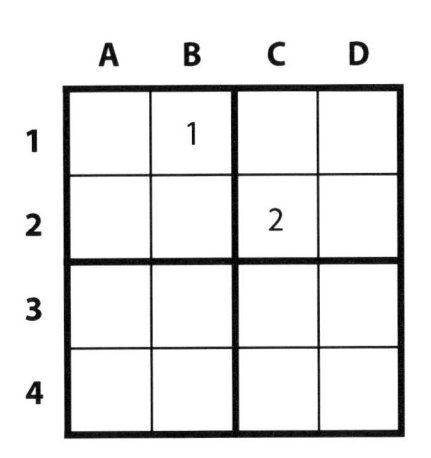

(1) 1

(2) 9

(3) $\dfrac{6\sqrt{10}}{2}$

(4) $\dfrac{3\sqrt{5}}{5}$

(4) $3 - \sqrt{3}$

	A	B	C	D
1		1		
2			2	
3				
4				

Radicals   37

OTHER ROOTS AND NEGATIVE EXPONENTS

$\sqrt{25}$ asks the question: what number multiplied by itself equals 25? The answer is $\sqrt{25} = 5$.

$\sqrt[3]{8}$ asks the question: what number times itself three times equals 8? The answer is $\sqrt[3]{8} = 2$, or $2 \cdot 2 \cdot 2 = 8$.

Also, a negative exponent implies a fraction. By definition, $x^{-a} = \dfrac{1}{x^a}$.

For example, $2^{-2} = \dfrac{1}{2^2} = \dfrac{1}{4}$.

Directions: Simplify the following. Find your answer in the box on the right. Place the numbers in the appropriate cells of the Sudoku grid, and then solve the puzzle.

(A3) $\sqrt[3]{125}$

(A6) 3^{-3}

(B2) $\sqrt[3]{1000}$

(B3) 4^{-2}

(C4) $\sqrt[3]{64x^6}$

(D1) $(xy)^{-3}$

(D3) $\sqrt[4]{16}$

(E4) 5^{-3}

(F1) $\sqrt[5]{32x^5}$

(F4) $(-7x^2)^{-3}$

(1) 10

(1) 2

(2) $4x^2$

(3) $\dfrac{1}{16}$

(3) $\dfrac{1}{27}$

(3) $\dfrac{1}{125}$

(4) $2x$

(5) $\dfrac{1}{-343x^6}$

(5) $\dfrac{1}{x^3 y^3}$

(6) 5

	A	B	C	D	E	F
1						
2						
3						
4						
5				5		
6			1			

Name _____ Date _____ Puzzle #30

INTRODUCTION TO QUADRATIC EQUATIONS

A **quadratic equation** is an equation in which the highest power of x is x^2. There are various methods of solving quadratic equations, including the square root method, factoring, and by using the quadratic formula. Generally, quadratic equations have two solutions.

In simple quadratic equations with the variable only represented once, the square root method is probably the easiest method.

Take the equation $2x^2 = 18$ as an example. To solve this, isolate x^2 on one side of the equation. Then find the square root of each side of the equation:

$2x^2 = 18$
$\frac{2x^2}{2} = \frac{18}{2}$
$x^2 = 9$
$\sqrt{x^2} = \sqrt{9}$
$x = 3$ and $x = -3$

Directions: Solve the quadratic equations. Find your answer in the box on the right. Place the numbers in the appropriate cells of the Sudoku grid, and then solve the puzzle. Show all work!

(A4) $x^2 = 49$

(A6) $3x^2 = 12$

(B2) $5x^2 = 45$

(B3) $x^2 - 7 = -6$

(C3) $24 - x^2 = -1$

(D1) $-4x^2 - 8 = -72$

(D4) $-3x^2 + 8 = -100$

(E4) $9x^2 = 4$

(F1) $25x^2 + 1 = 10$

(F3) $-5x^2 + \frac{2}{5} = -\frac{2}{5}$

(1) 4, -4
(2) 7, -7
(3) 1, -1
(3) $\frac{2}{3}, -\frac{2}{3}$
(4) 2, -2
(4) 3, -3
(4) $\frac{2}{5}, -\frac{2}{5}$
(5) 5, -5
(6) $\frac{3}{5}, -\frac{3}{5}$
(6) 6, -6

	A	B	C	D	E	F
1						
2						
3						
4						
5					6	
6			3			

Quadratic Equations

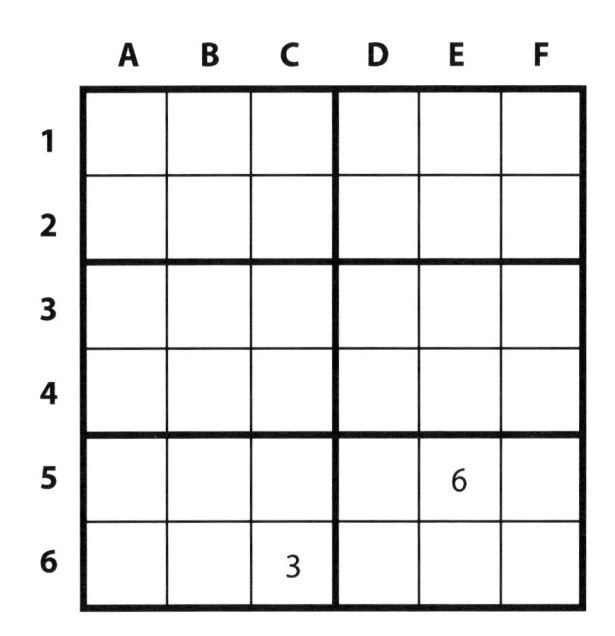

Name _____ Date _____ Puzzle #31

USING FACTORING TO SOLVE QUADRATIC EQUATIONS

The main reason for learning to factor polynomials is to use the skill as a tool in solving quadratic equations. First, write the quadratic equation in standard form. *The standard form for a quadratic equation is $ax^2 + bx + c = 0$.* Next, completely factor the left side of the equation. Then set each factor equal to zero, and solve each of them.

Here is an example of how it's done:

1) Write equation in standard form. $x^2 - 2x - 15 = 0$
2) Factor. $(x + 3)(x - 5) = 0$
3) Set factors equal to zero. $x + 3 = 0$ and $x - 5 = 0$
4) Solve. $x = -3$ and $x = 5$

Directions: Solve each quadratic equation using the factoring method. Find your answer in the box on the right. Place the numbers in the appropriate cells of the Sudoku grid, and then solve the puzzle. Show all work!

(A2 and A3) $x^2 - 2x - 15 = 0$

(B4 and B6) $4x^2 = 9$

(C6 and C4) $x^2 + 16 = -8x$

(D1 and D4) $2x^3 + 22x = -24x^2$

(E1 and E3) $3x^2 - 5x = 2$

(F4 and F5) $x^3 = 36x$

(1 and 2) 6, -6, 0

(1 and 4) -11, -1, 0

(2 and 5) $\frac{3}{2}, -\frac{3}{2}$

(2 and 6) $-\frac{1}{3}, 2$

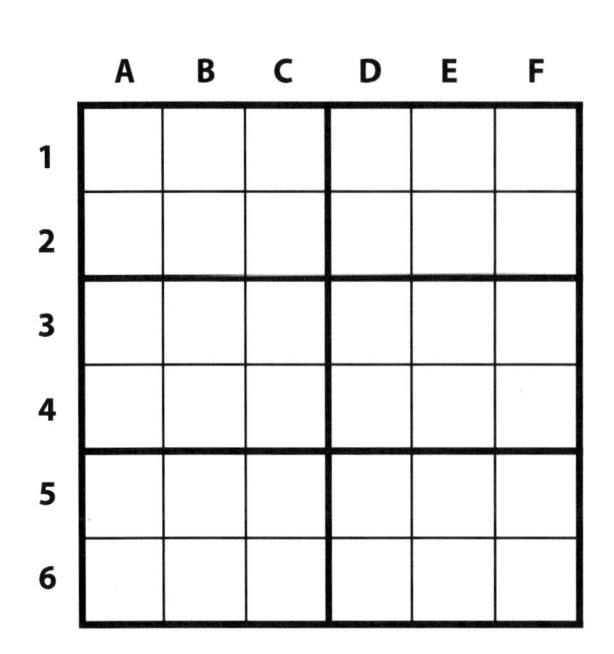

(3 and 6) -4, -4

(5 and 1) 5, -3

40

Quadratic Equations

Name _____ Date _____ Puzzle #32

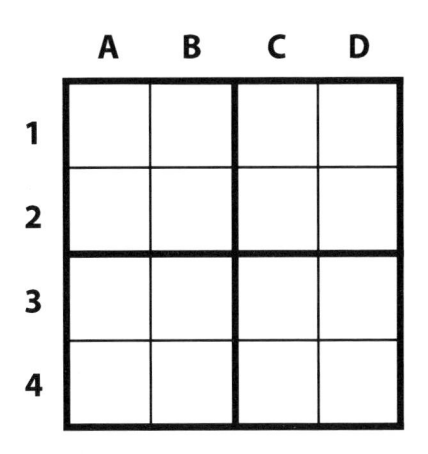

QUADRATIC FORMULA

The **quadratic formula**, $\dfrac{-b \pm \sqrt{b^2 - 4ac}}{2a}$, may be used to solve all quadratic equations. First, rearrange the equation into standard form, $ax^2 + bx + c = 0$. Next, substitute the values for *a*, *b*, and *c* into the equation. Then solve the equation.

For example:
$3x^2 - x - 2 = 0 \qquad a = 3, b = -1, c = -2$
Substitute the values into the quadratic formula, then solve as follows:

$x = \dfrac{-(-1) \pm \sqrt{(-1)^2 - 4(3)(-2)}}{2(3)}$

$x = \dfrac{1 \pm \sqrt{25}}{6} = \dfrac{1 \pm 5}{6}$

$x = \dfrac{1 + 5}{6}$ and $x = \dfrac{1 - 5}{6}$

$x = 1$ and $x = -\dfrac{2}{3}$

Directions: Solve the following quadratic equations using the quadratic formula. Find your answer in the box on the right. Place the numbers in the appropriate cells of the Sudoku grid, and then solve the puzzle. Show all work!

(A2 and B2)
$2x^2 + 5x = 3$

(C3 and D3)
$3x^2 - 2 = -5x$

(1 and 1) 4, -3

(2 and 1) $\dfrac{1}{2}$, -3

(3 and 1) $\dfrac{1}{3}$, -2

(4 and 2) 2, $-\dfrac{2}{3}$

Quadratic Equations

41

Name _____   Date _____   Puzzle #33

APPLICATION OF QUADRATIC EQUATIONS

Directions: Use a method of solving quadratic equations to find the answers to the two word problems below. Place the most reasonable answer in corresponding cells of the Sudoku grid, and then solve the puzzle. Show all work!

(A) Sade receives an allowance of $5 per week. One day she compared her allowance with that of her little brother Jorge. Sade determined that her allowance is equal to Jorge's allowance squared plus twice Jorge's allowance minus three dollars. How much is Jorge's allowance?

(B) During his physical exam for JV football, Andre learned that he now weighs 143 pounds. Andre calculated that he now weighs three times the square of his birth weight plus 26 times his birth weight minus nine pounds. How much did Andre weigh at birth?

	A	B	C	D
1		B	A	
2		A		B
3	A		B	
4	B			A

42   *Quadratic Equations*

ANSWER KEY

Puzzle #1

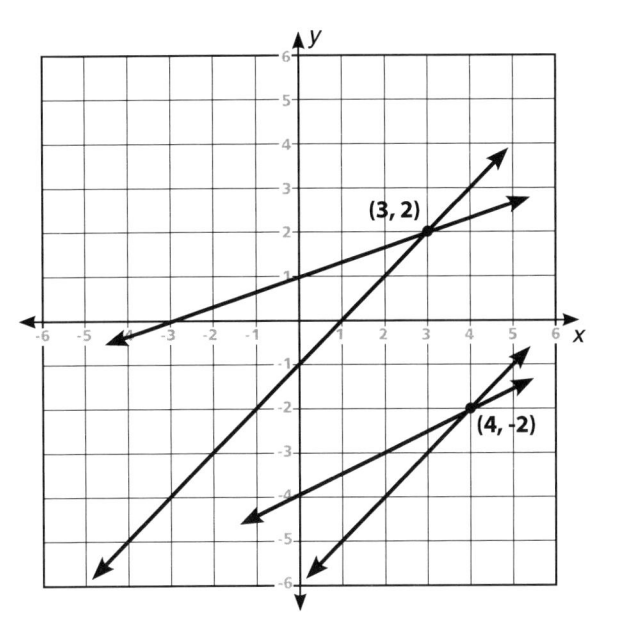

	A	B	C	D
1	4	2	3	1
2	3	1	4	-2
3	2	3	1	4
4	1	4	2	3

Puzzle #2

	A	B	C	D	E	F
1	2	6	3	5	1	4
2	1	5	4	3	6	2
3	4	2	5	1	3	6
4	3	1	6	2	4	5
5	-6	-3	2	-4	5	1
6	5	4	1	6	2	3

Puzzle #3

	A	B	C	D	E	F
1	3	1	4	2	6	5
2	2	5	6	4	1	3
3	1	4	-5	6	3	2
4	-6	2	3	1	5	4
5	4	3	1	5	2	6
6	5	6	2	3	-4	1

Puzzle #4

	A	B	C	D	E	F
1	1	2	5	3	4	6
2	3	6	4	2	5	1
3	2	5	-6	1	3	4
4	4	3	1	6	2	5
5	6	4	3	5	1	2
6	5	1	2	4	6	-3

Puzzle #5

	A	B	C	D	E	F
1	1	6	2	3	5	4
2	5	3	4	1	2	6
3	4	5	1	2	6	3
4	6	2	3	5	4	1
5	2	1	6	4	3	5
6	3	4	5	6	1	2

Puzzle #6

	A	B	C	D
1	1	4	2	3
2	3	2	4	1
3	2	3	1	4
4	4	1	3	2

Puzzle #7

	A	B	C	D	E	F
1	3	4	1	5	2	6
2	6	2	5	3	1	4
3	4	1	6	2	3	5
4	2	5	3	4	6	1
5	1	3	4	6	5	2
6	5	6	2	1	4	3

Puzzle #8

	A	B	C	D	E	F
1	5	3	2	6	4	-1
2	6	-1	4	3	5	2
3	-1	5	6	2	3	4
4	2	4	3	-1	6	5
5	3	2	5	4	-1	6
6	4	6	-1	5	2	3

Puzzle #9

	A	B	C	D	E	F
1	4	3	6	2	1	5
2	2	5	1	3	4	6
3	1	4	3	6	5	2
4	6	2	5	1	3	4
5	3	6	4	5	2	1
6	5	1	2	4	6	3

Puzzle #10

	A	B	C	D	E	F
1	1	6	2	5	4	3
2	5	3	4	6	1	2
3	2	1	5	4	3	6
4	3	4	6	1	2	5
5	6	2	1	3	5	4
6	4	5	3	2	6	1

Puzzle #11

	A	B	C	D	E	F
1	6	3	1	5	4	2
2	4	5	2	1	3	6
3	2	4	3	6	1	5
4	1	6	5	4	2	3
5	3	1	6	2	5	4
6	5	2	4	3	6	1

Puzzle #12

	A	B	C	D	E	F	G	H	I
1	9	5	1	3	8	2	4	7	6
2	6	7	8	5	4	9	1	3	2
3	2	3	4	1	6	7	5	9	8
4	4	9	7	8	2	6	3	1	5
5	8	1	5	7	9	3	2	6	4
6	3	2	6	4	1	5	7	8	9
7	1	4	2	6	7	8	9	5	3
8	5	6	9	2	3	1	8	4	7
9	7	8	3	9	5	4	6	2	1

Puzzle #13

	A	B	C	D	E	F
1	6	1	3	2	4	5
2	5	2	4	1	3	6
3	3	6	2	5	1	4
4	4	5	1	6	2	3
5	1	4	6	3	5	2
6	2	3	5	4	6	1

Puzzle #14

	A	B	C	D
1	2	1	3	4
2	4	3	1	2
3	1	4	2	3
4	3	2	4	1

Puzzle #15

	A	B	C	D
1	2	1	4	3
2	4	3	1	2
3	1	2	3	4
4	3	4	2	1

Puzzle #16

	A	B	C	D	E	F
1	2	1	3	4	6	5
2	6	5	4	3	2	1
3	4	2	5	1	3	6
4	3	6	1	5	4	2
5	5	3	2	6	1	4
6	1	4	6	2	5	3

Puzzle #17

	A	B	C	D	E	F
1	4	2	6	5	1	3
2	1	3	5	6	4	2
3	2	5	1	4	3	6
4	3	6	4	1	2	5
5	6	1	2	3	5	4
6	5	4	3	2	6	1

Puzzle #18

	A	B	C	D	E	F	G	H	I
1	4	2	9	3	7	1	8	5	6
2	6	5	8	4	2	9	7	1	3
3	1	7	3	5	6	8	2	9	4
4	2	8	6	9	5	7	3	4	1
5	7	9	1	2	4	3	5	6	8
6	5	3	4	1	8	6	9	7	2
7	3	6	2	7	9	4	1	8	5
8	8	1	7	6	3	5	4	2	9
9	9	4	5	8	1	2	6	3	7

Puzzle #19

	A	B	C	D	E	F
1	6	4	2	5	3	1
2	3	1	5	2	6	4
3	5	6	1	4	2	3
4	2	3	4	1	5	6
5	1	2	6	3	4	5
6	4	5	3	6	1	2

Puzzle #20

	A	B	C	D	E	F	G	H	I
1	3	7	9	4	6	1	5	2	8
2	8	1	5	9	2	7	4	3	6
3	6	2	4	8	3	5	7	1	9
4	7	5	1	6	8	3	9	4	2
5	2	8	6	5	9	4	1	7	3
6	9	4	3	7	1	2	8	6	5
7	1	6	8	2	7	9	3	5	4
8	4	3	2	1	5	8	6	9	7
9	5	9	7	3	4	6	2	8	1

Puzzle #21

	A	B	C	D	E	F
1	6	3	2	5	1	4
2	1	4	5	2	6	3
3	4	2	6	1	3	5
4	3	5	1	6	4	2
5	2	1	3	4	5	6
6	5	6	4	3	2	1

Puzzle #22

	A	B	C	D
1	1	2	3	4
2	3	4	1	2
3	4	1	2	3
4	2	3	4	1

Puzzle #23

	A	B	C	D	E	F
1	3	6	2	1	5	4
2	5	4	1	2	3	6
3	6	1	5	3	4	2
4	4	2	3	5	6	1
5	2	5	6	4	1	3
6	1	3	4	6	2	5

Puzzle #24

	A	B	C	D	E	F
1	4	3	6	5	2	1
2	2	1	5	6	4	3
3	3	5	2	4	1	6
4	1	6	4	2	3	5
5	6	2	3	1	5	4
6	5	4	1	3	6	2

Puzzle #25

	A	B	C	D	E	F
1	5	4	3	1	6	2
2	6	2	1	3	5	4
3	4	1	6	5	2	3
4	2	3	5	6	4	1
5	3	6	4	2	1	5
6	1	5	2	4	3	6

Puzzle #26

	A	B	C	D	E	F
1	3	1	2	4	6	5
2	6	5	4	2	3	1
3	1	4	6	3	5	2
4	5	2	3	6	1	4
5	2	6	1	5	4	3
6	4	3	5	1	2	6

Puzzle #27

	A	B	C	D	E	F
1	2	5	3	1	4	6
2	4	6	1	3	2	5
3	1	2	6	5	3	4
4	3	4	5	6	1	2
5	6	3	2	4	5	1
6	5	1	4	2	6	3

Puzzle #28

	A	B	C	D
1	2	1	4	3
2	4	3	2	1
3	1	2	3	4
4	3	4	1	2

49

Puzzle #29

	A	B	C	D	E	F
1	2	6	3	5	1	4
2	5	1	4	2	6	3
3	6	3	5	1	4	2
4	1	4	2	6	3	5
5	4	2	6	3	5	1
6	3	5	1	4	2	6

Puzzle #30

	A	B	C	D	E	F
1	3	5	2	1	4	6
2	1	4	6	3	5	2
3	6	3	5	2	1	4
4	2	1	4	6	3	5
5	5	2	1	4	6	3
6	4	6	3	5	2	1

Puzzle #31

	A	B	C	D	E	F
1	6	3	4	1	2	5
2	5	1	2	3	4	6
3	1	4	5	2	6	3
4	3	2	6	4	5	1
5	4	6	1	5	3	2
6	2	5	3	6	1	4

Puzzle #32

	A	B	C	D
1	3	4	1	2
2	2	1	4	3
3	4	2	3	1
4	1	3	2	4

Puzzle #33

	A	B	C	D
1	1	4	2	3
2	3	2	1	4
3	2	3	4	1
4	4	1	3	2

NOTES

REFERENCES

Adelman, C. *Answers in the tool box: Academic intensity, attendance patterns, and bachelor's degree attainment.* Washington, DC: U.S. Department of Education, Office of Educational Research and Improvement, 1999.

Business Higher Education Forum. *A Commitment to America's Future: Responding to the Crisis in Mathematics and Science Education - The Main Report.* Washington, DC, 2005.
http://www.bhef.com/www/publications/documents/commitment_future_05.pdf

Evan, A., Gray, T., & Olchefske, J. *The gateway to student success in mathematics and science.* Washington, DC: American Institutes for Research, 2006.

National Mathematics Advisory Panel, *Final Report.* Washington, DC, 2008.

Williams, Tony. *100 Algebra Workouts and Practical Teaching Tips.* Dayton: Teaching and Learning Company, 2008.

Williams, Tony. *100 Math Workouts and Practical Teaching Tips.* Dayton: Teaching and Learning Company, 2008.